FOREWORD

We hope you enjoy this coloring book! You will learn about and their country, Madagascar. We'd also like to introduce you to the nonprofit organization behind this coloring book: Green Again Madagascar.

Madagascar is a hotspot of biodiversity. Huge numbers of species of plants and animals (such as lemurs) are found nowhere else on Earth. However, the rainforest is disappearing as habitat is lost to slash and burn agriculture, an invasive fern, and fire. Green Again Restoration (better known as Green Again Madagascar) works to create forest plantations that can withstand the invasive ferns and the possibility of fire. This reforestation not only counteracts the loss of native forest and biodiversity, it helps local villagers, who are hired and trained to plant trees, take measurements and record data.

With an adult literacy rate in Madagascar of approximately 64.7% (much less in rural areas), several nonprofit organizations advised Green Again to work only with formally educated Malagasy people. But Matt Hill, our founder, was interested in working with rural farmers, regardless of formal education because they would have a deep appreciation and understanding of the land. Green Again believes that there are plenty of smart, capable people in the countryside who just never had an opportunity to get a good education.

Working with local landowners is a key to our success. We plant trees with cultural or economic value to landowners, such as species their grandfathers crafted canoes out of, or trees with medicinal uses. Local farmers make the decisions about what trees to plant and why they need those particular species. Family members and neighbors are all involved in preparing the land, planting, and weeding and Green Again provides per diem income for their work. Some per diem workers go on to become involved in Green Again operations, acquiring income stability and transferable skills in ecology, restoration, and team-management.

Another core part of our work is making scientific observations on the growth and survival of each tree that we plant. The data we gather is used for continual improvement of our techniques. And the data is managed by the data guru, Matt Hill, founder of Green Again. In his previous life, Matt worked on Wall Street as a quantitative analyst (a.k.a. a "quant") for stock portfolios worth billions of dollars. Now, at the helm of a forest conservation nonprofit in Madagascar, he finds himself applying the same data-driven approach he used in his former career to restore the island's imperiled forests.

Green Again is a 501(c)3 non-profit
EIN 83-3651734
https://www.greenagainmadagascar.org/
Follow us on Facebook!

Please see the back of the "Title page" for some suggestions for how your family can use this coloring book. And note that this is only Book 1! We have another in the works with great art by high school students, professional artists, and Malagasy collaborators.

MYSTERIOUS & MAGNIFICENT RAINFOREST ANIMALS OF MADAGASCAR
Coloring Book #1

ORGANIZATION OF THIS BOOK:

- **Suggestions for the use of this book** on the back of this page
- **Illustrations 1-37** with "fun facts" on the back of each page
- **Animal Facts** - seven pages of more detailed animal information - a great opportunity to learn about these species, Madagascar, and what conditions support the biodiversity of the country
- **Artist credits** - a brief introduction to the amazing illustrators involved in this project
- **More Green Again history** and organization contact information on the inside back cover. We hope you will follow us on Facebook!

A Special Thanks to Patrick Moran!
Thanks so much for the marvelous cover art
and book assembly by
MORAN
www.artofmoran.com
Artist/Animator

Cover art by MORAN with Roland & Pascal

Suggestions for use of this book

This book was created by artists from around the world - high school students, Peace Corps workers, professional artists, Malagasy residents involved in rainforest restoration. The styles are different and we purposely included art that a 5 year old could proudly color - and other pages that an adult needing something lovely to color creatively could do to relax. We hope the whole family may find something they treasure in this book.

The header and footer on the back of each coloring page include some fun and interesting facts about the species you will be coloring. We hope this will inspire you to learn more about the rainforest and its residents.

At the back of the book we have several pages of more extensive information about each species. Both the English name and the scientific name of each species are included, although the list is alphabetized by the English name.

There is credits page by artist in case you fall in love with the style of one artist in particular - or want to find the art of a child, grandchild, or friend. We will also include brief information as available from each artist.

If you are interested in scientific research papers on our work please leave a message on our website. Our reliance on science is one of the things that makes us different from other tree planting groups. We are actually in the business of growing trees, not just planting them. We pay attention to what conditions allow trees to survive drought, fire, cyclone, competition with invasive species.

If you are interested in learning more about this organization and our founder, Matt Hill, please visit our Facebook page - Green Again Madagascar

Or explore our webpage - https://www.greenagainmadagascar.org/

Note that our actual organizational name is Green Again Restoration, and that is the name of our 501c(3). We do business as Green Again Madagascar.

Green Again Madagascar
www.GreenAgainMadagascar.org

No.1

Mouse Lemur
Microcebus tavaratra

Amanda Albert
Florida, USA

This mouse lemur is found in northern Madagascar and like all mouse lemurs it is nocturnal. Mouse lemurs have a combined head, body and tail length of less than 11 inches, making them the smallest primates.

In 1992, there were two known mouse lemur species; by 2016, there were 24! The number of species is uncertain and is increasing as more animals are discovered and studied.

The aye-aye is the world's largest nocturnal primate.
This unusual looking creature is very small, around 14-17 inches

It has an unusual method of finding food: it taps on trees to find grubs,
then gnaws holes in the wood using its forward-slanting incisors to create a small hole
in which it inserts its narrow middle finger to pull the grubs out.

Green Again Madagascar
www.GreenAgainMadagascar.org

No.3

giraffe weevil
trachelophorus giraffa

Taylor Palacino
Virginia, USA

This weevil gets its name from an extended neck, much like that of a giraffe. Giraffe weevils live in forests, spending almost their entire lives on a tree known as the "giraffe beetle tree".

Giraffe weevils cannot bite or sting and, therefore, are not dangerous to humans. This species was only discovered in 2008, so not a lot is known about them yet!

Green Again Madagascar
www.GreenAgainMadagascar.org

No.4

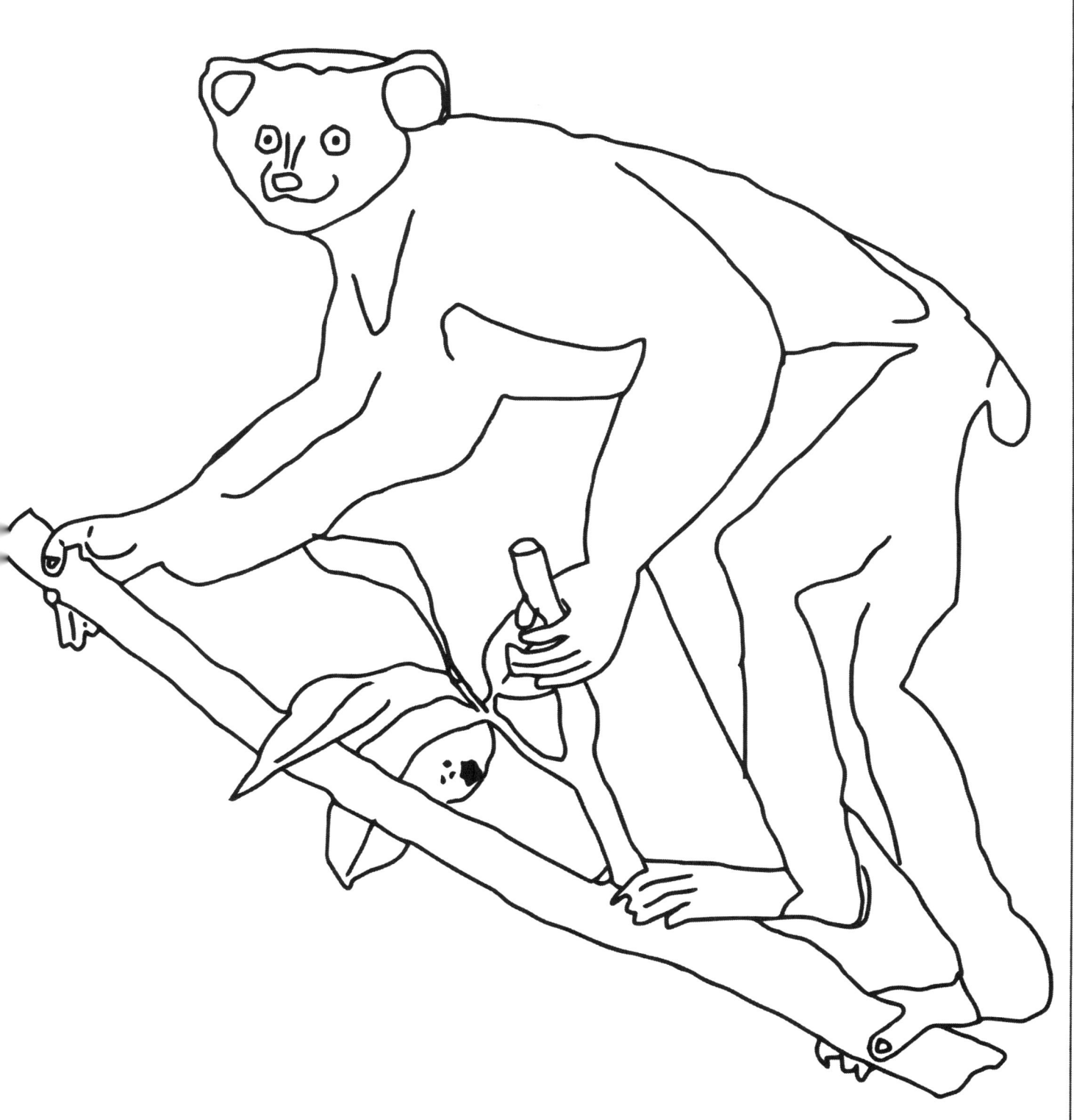

ndril

ndri-indri

Stephan Gray

Florida, USA

The Indri is one of the largest living lemurs, with a length of about 25–28 inches and a weight of 13-21 pounds. It has long, muscular legs which it uses to propel itself from trunk to trunk

Groups of indri are quite vocal, communicating with other groups by singing, roaring and other vocalizations. Before singing, the indri move to treetops, which allows them to be heard up to almost 2.5 miles away.

Green Again Madagascar
www.GreenAgainMadagascar.org
No.5

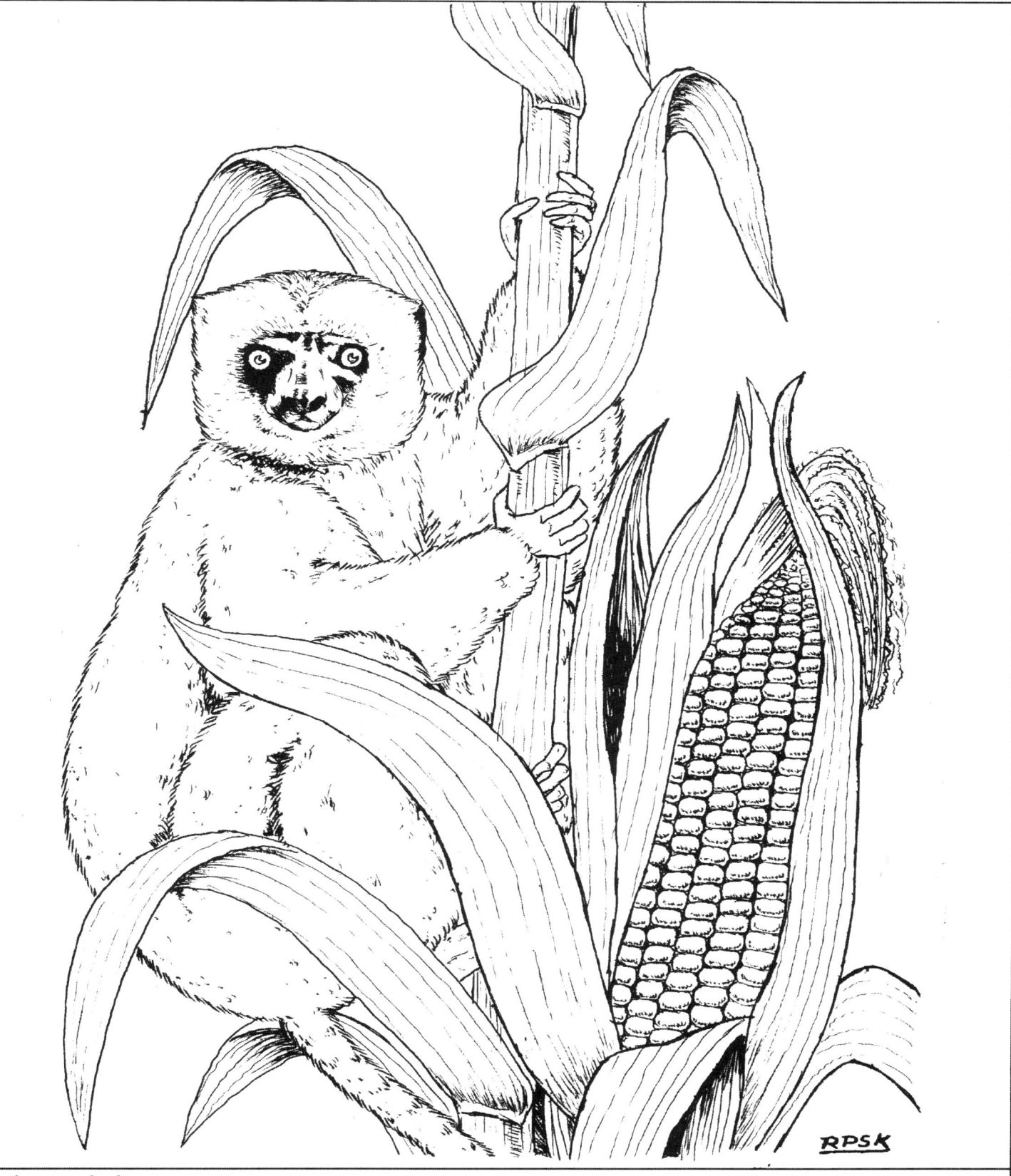

Silky sifaka — Roland & Pascal
Propithecus candidus — Moramanga, Madagascar

This large lemur is characterized by long, silky, white fur. It is only found within a few protected areas in the rainforests of northeastern Madagascar. and is one of the rarest mammals on Earth.

The silky sifaka lives in groups of two to nine individuals. Like other sifakas, it eats mainly leaves and seeds, but also fruit, flowers, and even soil on occasion.

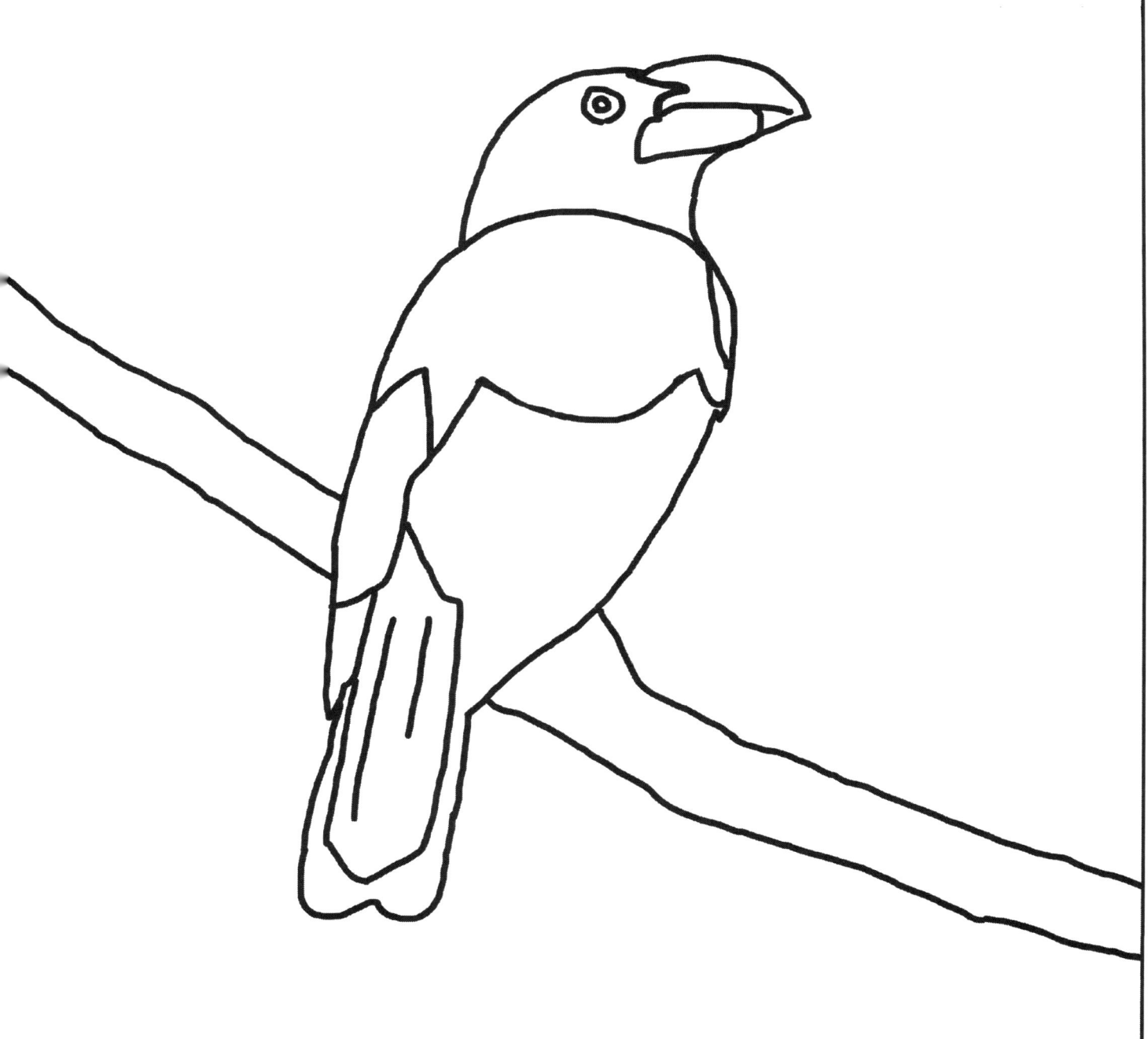

The helmet vanga is a distinctive-looking bird.
It is mainly blue-black, with rufous wings and a huge arched blue bill,
which is 2.0 inches long and 1.2 inches deep.

This bird is very secretive and hard to find, often sitting motionless in a tree for long periods.
Its large bill allows it to feed on large prey such as insects, amphibians and lizards.

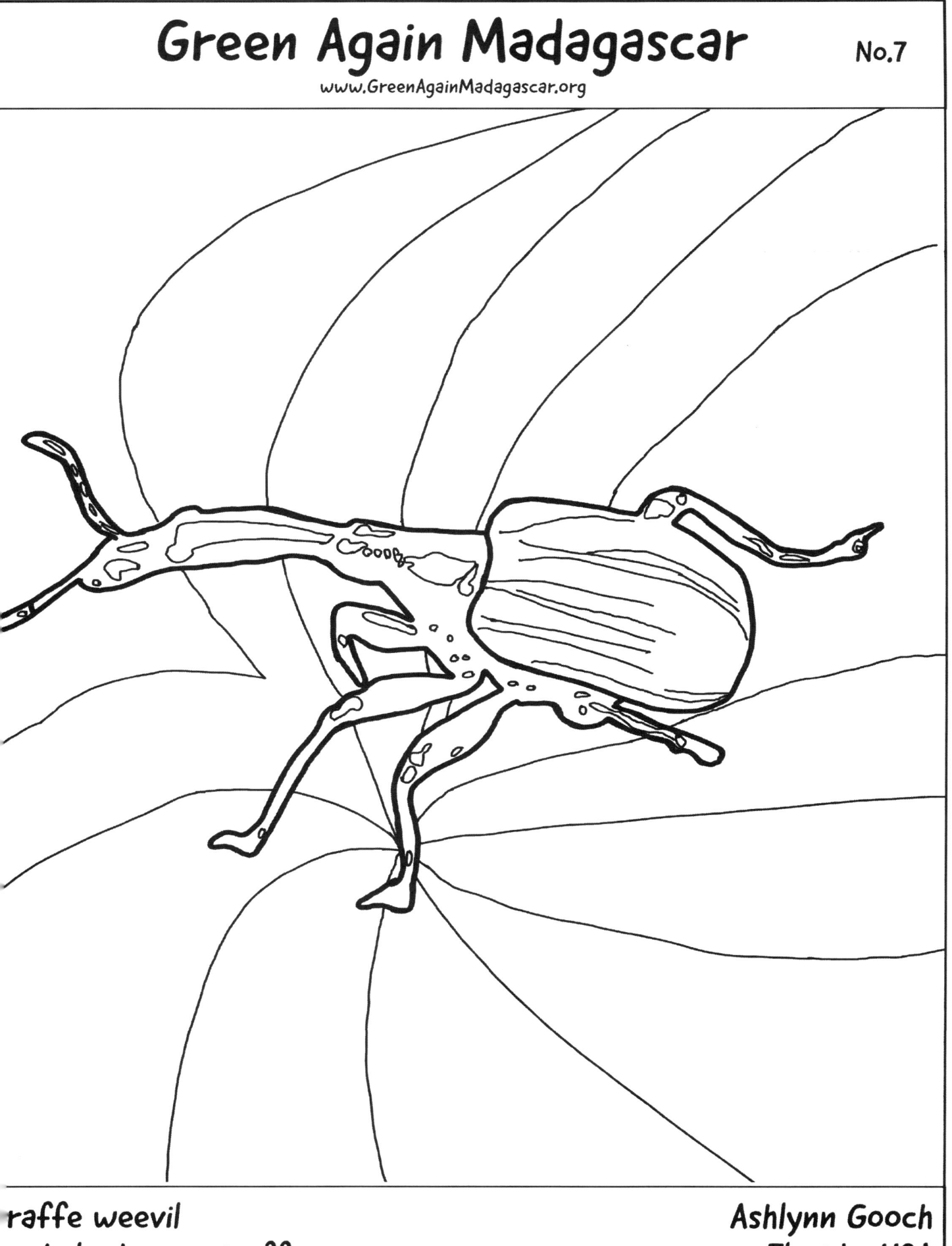

This weevil gets its name from an extended neck,
much like that of a giraffe. Giraffe weevils live in forests,
spending almost their entire lives on a tree known as the "giraffe beetle tree".

Giraffe weevils cannot bite or sting and, therefore, are not dangerous to humans.
This species was only discovered in 2008, so not a lot is known about them yet!

This bat is named for its suction cup like feet that allow for it to grasp onto just about any surface. They roost inside the rolled leaves of a tree, using their suckers to attach themselves to the smooth surface.

The Madagascar sucker-footed bat is one of the few bat species that roosts with its head up rather than upside down, so the bat does not accidentally lose control of the adhesive pads on its feet while it is sleeping!

Green Again Madagascar
www.GreenAgainMadagascar.org

No.9

raffe weevil
achelophorus giraffa

Lea Ritson
Vohitsara, Madagascar

This weevil gets its name from an extended neck,
much like that of a giraffe. Giraffe weevils live in forests,
spending almost their entire lives on a tree known as the "giraffe beetle tree".

Giraffe weevils cannot bite or sting and, therefore, are not dangerous to humans.
This species was only discovered in 2008, so not a lot is known about them yet!

Green Again Madagascar

www.GreenAgainMadagascar.org

No. 10

Broad-striped vontsira
Galidictis fasciata

Lea Ritson
Vohitsara, Madagascar

The broad-striped Malagasy mongoose is a forest-dweller on the eastern side of the island, finding its home in lowland forests. Their primary prey is small rodents and they are most active at night and usually like company.

Broad-striped mongooses have nimble, low-to-the-ground bodies. They have short legs and long bushy tails. Their heads are long and slender.

Green Again Madagascar
www.GreenAgainMadagascar.org

No.11

Panther chameleon
Furcifer pardalis

Roland & Pascal
Moramanga, Madagascar

The panther chameleon has five toes fused into a group of two and a group of three, giving the foot a tongs-like appearance. These specialized feet allow the panther chameleon a tight grip on narrow branches

Their upper and lower eyelids are joined, with only a pinhole large enough for the pupil to see through. They can rotate and focus their eyes separately to observe two different objects simultaneously!

Green Again Madagascar
www.GreenAgainMadagascar.org

No.12

eaf-nosed snake
angaha madagascariensis

Lea Ritson
Vohitsara, Madagascar

Commonly known as the Madagascar or Malagasy leaf-nosed snake.
It can grow up to 3 feet in length. Females have a flattened, leaf shaped snout.

Malagasy leaf-nosed snakes are generally calm and reluctant to bite unless provoked.
A bite causes severe pain in humans but is not deadly.

Green Again Madagascar

No.13

www.GreenAgainMadagascar.org

eaf-tailed gecko
roplatus phantasticus

Chris Chanaud
California, USA

Leaf-tailed geckos have large eyes and broad flat, leaf-like tails.
Because they have no eyelids, just a transparent covering over their eyes,
they use their long, mobile tongues to wipe away any dust or debris that gets into their eyes

Predators hardly notice it among the branches and leaves due to its excellent camouflage.
The leaf-tailed gecko is even able to flatten the body to reduce its shadow,
and can shed its tail to escape from danger

Green Again Madagascar
www.GreenAgainMadagascar.org

No. 14

eaf-nosed snake
angaha madagascariensis

Isabella Choice
Florida, USA

Commonly known as the Madagascar or Malagasy leaf-nosed snake.
It can grow up to 3 feet in length. Females have a flattened, leaf shaped snout.

Malagasy leaf-nosed snakes are generally calm and reluctant to bite unless provoked.
A bite causes severe pain in humans but is not deadly.

Green Again Madagascar
www.GreenAgainMadagascar.org

No.15

Ring-tailed lemur
Lemur catta

Roland & Pascal
Moramanga, Madagascar

The ring-tailed lemur is a large primate and the most recognized lemur due to its long, black and white ringed tail. This lemur is highly social, living in groups of up to 30 individuals.

It is a female dominant society, a trait common among lemurs. Males perform a unique scent marking behavior called spur marking and will participate in stink fights by impregnating their tail with their scent and wafting it at opponents.

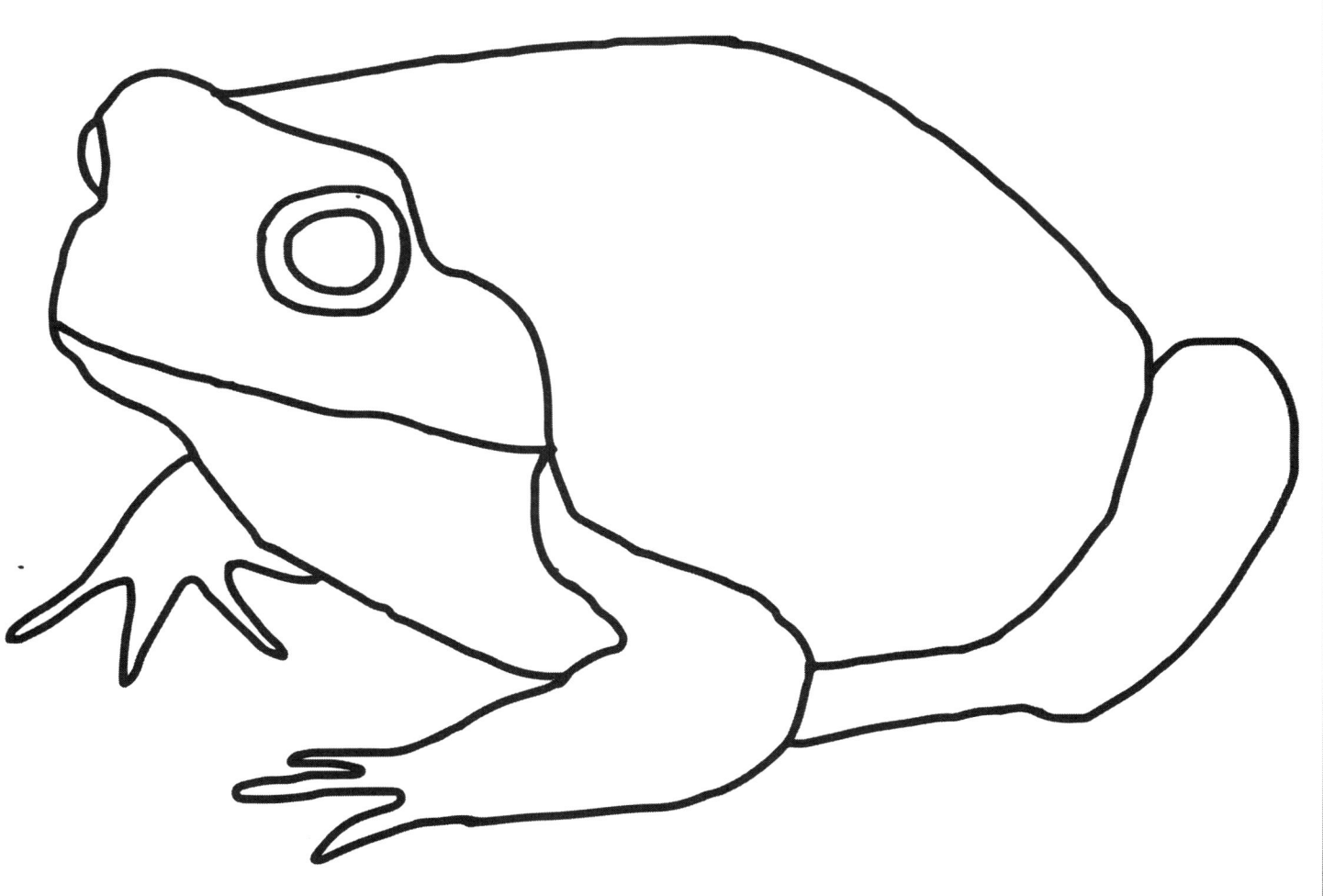

Tomato frogs are found in the northeast of Madagascar. Females are much larger than males, at about 4 inches long and about 8 ounces in weight.

Tomato frogs have a vibrant, orange-red color.
It is thought that the brilliant colors of the tomato frog act as a warning to potential predators that these frogs are toxic.

Green Again Madagascar

www.GreenAgainMadagascar.org

No.17

Blue coua
Coua caerulea

Lea Ritson
Vohitsara, Madagascar

Found in subtropical and tropical forests of northwestern and eastern areas of Madagascar, the blue coua is a cuckoo with deep blue feathers. It is an omnivore, eating insects, fruits and small reptiles.

Females lay a single white egg on a platform nest constructed of leaves and twigs on a tree branch.
Both adults share the incubation and feed the young with insects.

Green Again Madagascar

No.18

www.GreenAgainMadagascar.org

tomato frog
Dyscophus antongilii

Lea Ritson
Vohitsara, Madagascar

Tomato frogs are found in the northeast of Madagascar. Females are much larger than males, at about 4 inches long and about 8 ounces in weight.

Tomato frogs have a vibrant, orange-red color. It is thought that the brilliant colors of the tomato frog act as a warning to potential predators that these frogs are toxic.

Green Again Madagascar

No.19

www.GreenAgainMadagascar.org

Malagasy kingfisher
Corythornis vintsiodes

Lea Ritson
Vohitsara, Madagascar

This beautiful little bird's natural habitat is subtropical or tropical mangrove forests. It is one of only two kingfishers that live in Madagascar.

Malagasy kingfishers are fast fliers, skimming the water as their short, rounded wings whir. They perch near a river or stream watching for prey to swim into striking range.

Green Again Madagascar

No.20

www.GreenAgainMadagascar.org

Tomato frog
Dyscophus antongilii

Taylor Palacino
Virginia, USA

Tomato frogs are found in the northeast of Madagascar.
Females are much larger than males, at about 4 inches long
and about 8 ounces in weight.

Tomato frogs have a vibrant, orange-red color.
It is thought that the brilliant colors of the tomato frog
act as a warning to potential predators that these frogs are toxic.

_# Green Again Madagascar
www.GreenAgainMadagascar.org

No. 21

Panther chameleon
Furcifer pardalis

Stephanie Ribich
Florida, USA

The panther chameleon has five toes fused into a group of two and a group of three, giving the foot a tongs-like appearance. These specialized feet allow the panther chameleon a tight grip on narrow branches.

Their upper and lower eyelids are joined, with only a pinhole large enough for the pupil to see through. They can rotate and focus their eyes separately to observe two different objects simultaneously!

Green Again Madagascar
No.22

www.GreenAgainMadagascar.org

Malthe's green-eared chameleon
Calumma malthe

John-Henry Dickinson
Florida, USA

This chameleon is found in a few isolated pockets and only in the middle of rainforests in dense vegetation near small rivers where the humidity is high. The male has a funny looking nose appendage.

These chameleons spread their occipital lobes to look larger and more impressive, when they are stressed or being aggressive. They also hiss.

Green Again Madagascar
www.GreenAgainMadagascar.org

No.23

Helmet vanga
Euryceros prevostii

Roland & Pascal
Moramanga, Madagascar

The helmet vanga is a distinctive-looking bird. It is mainly blue-black, with rufous wings and a huge arched blue bill, which is 2.0 inches long and 1.2 inches deep.

This bird is very secretive and hard to find, often sitting motionless in a tree for long periods Its large bill allows it to feed on large prey such as insects, amphibians and lizards.

Green Again Madagascar

No.24

www.GreenAgainMadagascar.org

Malthe's green-eared chameleon
Calumma malthe

Lea Ritson
Vohitsara, Madagascar

This chameleon is found in a few isolated pockets and only in the middle of rainforests in dense vegetation near small rivers where the humidity is high. The male has a funny looking nose appendage.

These chameleons spread their occipital lobes to look larger and more impressive, when they are stressed or being aggressive. They also hiss.

Green Again Madagascar

No.25

www.GreenAgainMadagascar.org

Red-ruffed lemur
Varecia rubra

Nathan Sharkey
Florida, USA

Red ruffed lemurs are entirely vegetarian.
They eat some leaves, seeds, and nectar, but most of their diet consists of fruit.

It is one of the largest primates of Madagascar.
Ruffed lemurs are one of only a few primates that have litters of offspring.

Green Again Madagascar
www.GreenAgainMadagascar.org

No.26

Silky sifaka
Propithecus candidus

Chris Chanaud
California, USA

This large lemur is characterized by long, silky, white fur. It is only found within a few protected areas in the rainforests of northeastern Madagascar. and is one of the rarest mammals on Earth.

The silky sifaka lives in groups of two to nine individuals. Like other sifakas, it eats mainly leaves and seeds, but also fruit, flowers, and even soil on occasion.

Green Again Madagascar

www.GreenAgainMadagascar.org

No.27

Red-legged golden orb-weaver
Nephila inaurata

Chris Chanaud
California, USA

This spider is extremely large, notable for its strength and characteristic golden color. It has long, slender red legs and a highly elongated,
silvery and black abdomen which may have prominent yellow markings

The silky sifaka lives in groups of two to nine individuals.
Like other sifakas, it eats mainly leaves and seeds, but also fruit,
flowers, and even soil on occasion.

Green Again Madagascar
www.GreenAgainMadagascar.org

No.28

Comet moth
Agrema mittrei

Chris Chanaud
California, USA

This beautiful golden yellow moth is native to the rainforest.
The male has a wingspan of almost 8 inches and a tail length of 6 inches
making it one of the world's largest silk moths.

The female lays 120-170 eggs, and after hatching, the larvae feed on leaves before pupating.
The whitish-gray cocoon has holes to keep the pupa from drowning
in the daily rains of its natural habitat.

Green Again Madagascar

No. 29

www.GreenAgainMadagascar.org

Malagasy fruit bat
Eidolon dupreanum

Amanda Albert
Florida, USA

The Madagascan fruit bat is found both around the coastal plain and in the inland high plateau.
It mainly eats fruit but also feeds on the flowers of Eucalyptus trees.

A single offspring is produced each year. A young or baby of a Madagascan fruit bat is called a 'pup'. A Madagascan fruit bat group is called a 'colony or cloud'.

Green Again Madagascar
www.GreenAgainMadagascar.org

No.30

ndri
ndri-indri

Roland & Pascal
Moramanga, Madagascar

The Indri is one of the largest living lemurs, with a length of about 25–28 inches and a weight of 13-21 pounds. It has long, muscular legs which it uses to propel itself from trunk to trunk.

Groups of indri are quite vocal, communicating with other groups by singing, roaring and other vocalization. Before singing, the indri move to treetops, which allows them to be heard up to almost 2.5 miles away.

Green Again Madagascar
www.GreenAgainMadagascar.org

No. 31

Madagascar bright-eyed frog
Boophis madagascariensis

Emily Krupp
Fazihay, Madagascar

This 2.4 to 4 inch tree frog is one of Madagascar's largest frogs.
It is found in the rainforest of the highlands down to the east coast.

Females can lay 400 eggs or more!
The eggs are black and can be found in shallow, slow-moving water, in the shade.

Green Again Madagascar
www.GreenAgainMadagascar.org
No.32

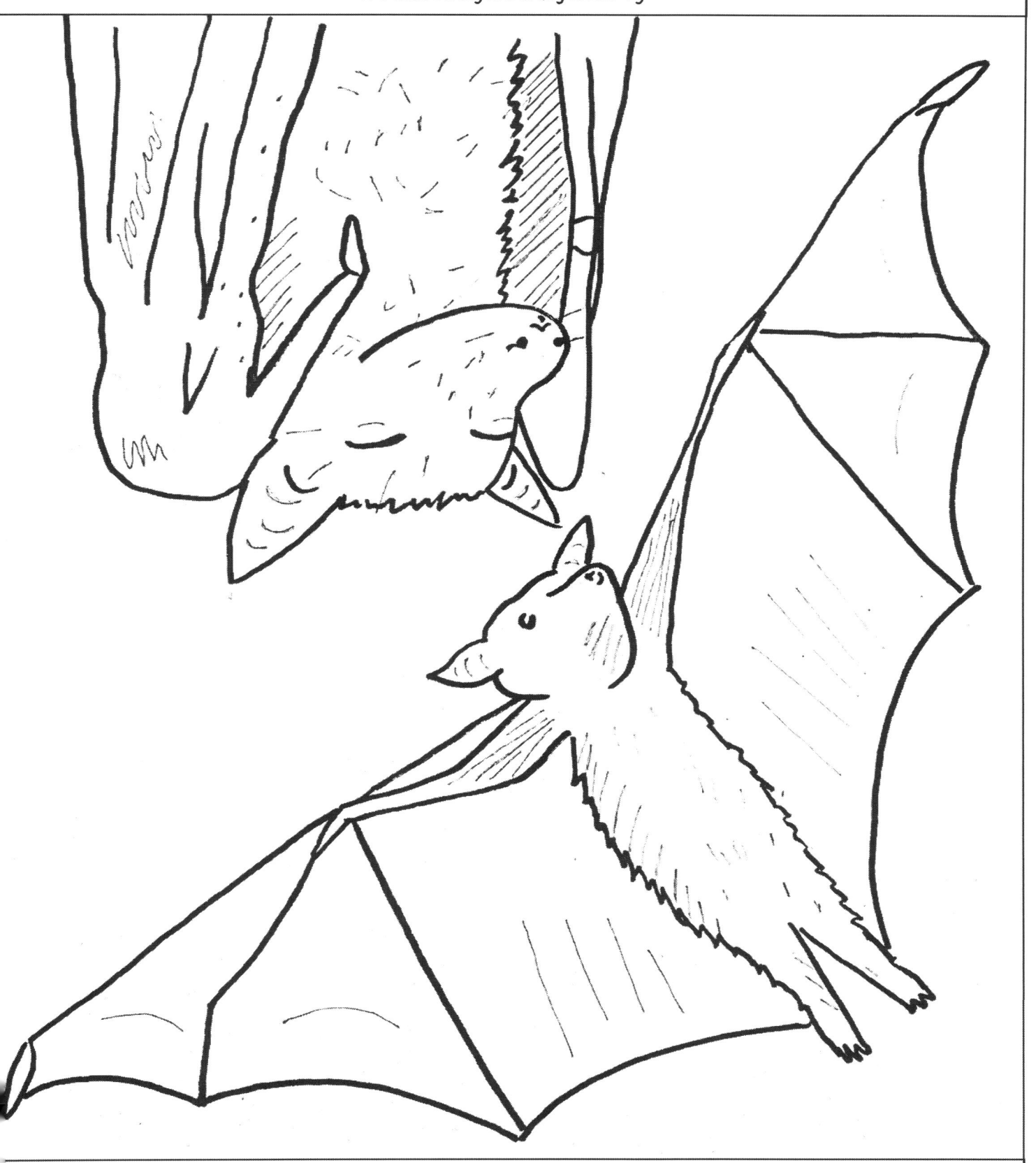

Malagasy fruit bat
Eidolon dupreanum

Emily Krupp
Fazihay, Madagascar

The Madagascan fruit bat is found both around the coastal plain
and in the inland high plateau.
It mainly eats fruit but also feeds on the flowers of Eucalyptus trees.

A single offspring is produced each year. A young or baby of a Madagascan fruit bat is called a 'pup'. A Madagascan fruit bat group is called a 'colony or cloud'.

Green Again Madagascar
www.GreenAgainMadagascar.org

No.33

Malagasy kingfisher
Corythornis vintsiodes

Emily Krupp
Fazihay, Madagascar

This beautiful little bird's natural habitat is subtropical or tropical mangrove forests. It is one of only two kingfishers that live in Madagascar.

Malagasy kingfishers are fast fliers, skimming the water as their short, rounded wings whir. They perch near a river or stream watching for prey to swim into striking range

Green Again Madagascar

No.34

www.GreenAgainMadagascar.org

Mouse lemur
Microcebus tavaratra

Emily Krupp
Fazihay, Madagascar

This mouse lemur is found in northern Madagascar and like all mouse lemurs it is nocturnal. Mouse lemurs have a combined head, body and tail length of less than 11 inches, making them the smallest primates.

In 1992, there were two known mouse lemur species; by 2016, there were 24! The number of species is uncertain and is increasing as more animals are discovered and studied.

Green Again Madagascar
www.GreenAgainMadagascar.org

No. 35

Panther chameleon
Furcifer pardalis

Emily Krupp
Fazihay, Madagascar

Leaf-tailed geckos have large eyes and broad flat, leaf-like tails.
Because they have no eyelids, just a transparent covering over their eyes,
they use their long, mobile tongues to wipe away any dust or debris that gets into their eyes.

Predators hardly notice it among the branches and leaves due to its excellent camouflage.
The leaf-tailed gecko is even able to flatten the body
to reduce its shadow, and can shed its tail to escape from danger.

Green Again Madagascar
No.36
www.GreenAgainMadagascar.org

Leaf-tailed gecko
Uroplatus phantasticus

Kat Culbertson
Ampasimbe, Madagascar

Leaf-tailed geckos have large eyes and broad flat, leaf-like tails.
Because they have no eyelids, just a transparent covering over their eyes,
they use their long, mobile tongues to wipe away any dust or debris that gets into their eyes.

Predators hardly notice it among the branches and leaves due to its excellent camouflage.
The leaf-tailed gecko is even able to flatten the body
to reduce its shadow, and can shed its tail to escape from danger.

Green Again Madagascar

No.37

www.GreenAgainMadagascar.org

Madagascar bright-eyed frog
Boophis madagascariensis

Kat Culbertson
Ampasimbe, Madagascar

This 2.4 to 4 inch tree frog
is one of Madagascar's largest frogs.
It is found in the rainforest of the highlands down to the east coast.

Females can lay 400 eggs or more!
The eggs are black and can be found in shallow, slow-moving water, in the shade.

Book 1 - Animal Facts

Aye aye   *Daubentonia madagascariensis*
The aye aye is the world's largest nocturnal primate. This unusual looking creature is very small, around 14-17 inches. Its tail is much longer than its body and measures 24 inches. The aye aye has large golden-colored eyes, which glow in the dark.

It has an unusual method of finding food: it taps on trees to find grubs, then gnaws holes in the wood using its forward-slanting incisors to create a small hole in which it inserts its narrow middle finger to pull the grubs out. The aye-aye was thought to be extinct in 1933, but was rediscovered in 1957.

The aye aye barely comes down from trees. It builds several nests in its territory using leaves and twigs and frequently keeps changing nests in order to escape from predators. Its ears are very large and naked, and it is capable of acute hearing. It is listed as an endangered species due to habitat loss and a Malagasy belief that the aye aye is evil.

Blue coua   *Coua caerulea*
Found in subtropical and tropical forests of northwestern and eastern areas of Madagascar, the blue coua is a cuckoo with deep blue feathers. It is an omnivore, eating insects, fruits and small reptiles. The blue coua is an arboreal species. It hunts by inspecting the foliage for prey. It is active from treetops to undergrowth, but never on the ground.

Females lay a single white egg on a platform nest constructed of leaves and twigs on a tree branch. **Both adults share the incubation and feed the young with insects.** Like all cuckoos, they have large feet with a reversible third toe. Its average size is 48 to 50 cm (18.9 to 19.7 in) in length and 30 - 60 grams (1.1 - 2.1 ounces) in weight with the females slightly larger. They are not threatened at this time.

Broad-striped vontsira   *Galidictis fasciata*
The broad-striped Malagasy mongoose is a forest-dweller on the eastern side of the island, finding its home in lowland forests. Their primary prey is small rodents and they are most active at night and usually like company. In camera trap surveys, the species was recorded primarily hanging out in pairs. While some of its mongoose cousins are strong climbers and love hanging out in the trees, this species sticks to the forest floor.

Broad-striped mongooses have nimble, low-to-the-ground bodies. They are small to medium in size, comparable to American martens. They have short legs and long bushy tails. Their heads are long and slender. Its body has about five broad dark brown or black stripes and creamy-beige stripes. Their very distinctive tails are a creamy white. Their ears are small and are covered with short, fine fur.

It weighs between 500 and 800 grams (18-28 oz). The length of head and body is 320-340 mm (~ 13 in) and tail length is 280-300 mm (11-12 in). Females are slightly smaller and lighter than males. Feet have longer digits and claws than other mongooses. It is listed as vulnerable.

Comet moth   *Argema mittrei*
This beautiful golden yellow moth is native to the rain forests of Madagascar. The male has a wingspan of 20 cm (almost 8 inches) and a tail length of 15 cm (6 inches), making it one of the world's largest silk moths. The female lays between 120-170 eggs, and after hatching, the

larvae feed on leaves for about two months before pupating. The whitish-gray cocoon has holes to keep the pupa from drowning in the daily rains of its natural habitat. The adult moth cannot feed and only lives for 4 to 8 days. The comet moth is endangered due to habitat loss.

Eastern sucker-footed bat *Myzopoda aurita*
This sucker-footed bat is native to the primary and secondary rainforest of eastern Madagascar. This species of bat can also be found in agricultural lands as well as urban and surrounding areas. The bat has shown to be quite adaptable to new locations as long as there are sources of food, water, and shelter for the species to rely on. Sucker-footed bats feed largely on beetles and small moths.

This bat is named for its suction cup like feet that allow for it to grasp onto just about any surface. They roost inside the rolled leaves of a tree, using their suckers to attach themselves to the smooth surface. Despite the name, it is now known that the bats do not use suction to attach themselves to roost sites, but instead use a form of wet adhesion by secreting body fluid at their pads. The ankle and wrist pads of the bat are controlled by muscle contraction and allow the bat to separate the pads to reduce the adhesive effect. This allows the bats to climb with ease and to remove themselves from surfaces after sticking. The Madagascar sucker-footed bat is one of the few bat species that roosts with its head up rather than upside down, so the bat does not accidentally lose control of the adhesive pads while it is sleeping.

Fruit bat   *Eidolon dupreanum*
The Madagascan fruit bat is found both around the coastal plain and in the inland high plateau. It mainly eats fruit but also feeds on the flowers of *Eucalyptus* trees. It was found to fly as far as 5 km (3.1 mi) to reach favored trees. This helps with dispersal of seed, and it has been shown that seed germination was enhanced by passage through the gut of the fruit bat. At times of year when fruit is not available, this bat feeds on the nectar of the flowers of some trees, and it is believed to pollinate those trees.

A single offspring is produced each year. This slow reproductive rate makes this bat susceptible to over-hunting. A young or baby of a Madagascan fruit bat is called a 'pup'. A Madagascan fruit bat group is called a 'colony or cloud'. It is listed as vulnerable, because it is hunted for its meat.

Giraffe weevil *Trachelophorus giraffa*
Giraffe weevils live in forests, spending almost their entire lives on a tree known as the "giraffe beetle tree". They are herbivorous, feeding on the leaves of that tree.

This weevil gets its name from an extended neck, much like that of a giraffe. The neck of the male is typically 2 to 3 times the length of that of the female. Most of the body is black with distinctive red covering the flying wings. The total body length of the males is just under one inch (2.5 cm). The extended neck is an adaptation that assists in nest building and fighting.

Females will roll a leaf up and lay a single egg inside the tube, snipping it off to fall onto the forest floor. The leaf then provides the larva with food in its first days of life.
Giraffe weevils cannot bite or sting and, therefore, are not dangerous to humans. This species was only discovered in 2008, so not a lot is known about them yet!

Helmet vanga   *Euryceros prevostii*
The helmet vanga is a distinctive-looking bird. It is mainly blue-black, with rufous wings and a huge arched blue bill. Both sexes look alike. It is restricted to lowland and lower mountain

rainforests of northeastern Madagascar. Its diet is composed of invertebrates, predominantly insects. It measures 28-31 cm (11–12 in) in length, and weighs 84-114 g (3.0–4.0 oz). The most distinctive feature is its massive hooked bill, which is 51 mm (2.0 in) long and 30 mm (1.2 in) deep.

The species is restricted to undisturbed humid rainforests, and this habitat is increasingly being cleared for agriculture and forestry. Their population, between 600-15,000 birds, is becoming increasingly fragmented. The helmet vanga is threatened with extinction due to habitat loss. It is believed that much of their remaining habitat will be lost in 50 years due to climate change.

Indri   *Indri indri*

The Indri inhabits the lowland and montane forests along the eastern coast of Madagascar. It is one of the largest living lemurs, with a length of about 64–72 cm (25–28 in) and a weight of 6-9.5 kg (13-21 lb). It has black and white fur and maintains an upright posture when climbing or clinging. It has long, muscular legs which it uses to propel itself from trunk to trunk. Its large greenish eyes and black face are framed by round, fuzzy ears that some say give it the appearance of a teddy bear. Unlike any other living lemur, the indri has only a rudimentary tail. It is monogamous and lives in small family groups. It is herbivorous, feeding mainly on leaves but also seeds, fruits, and flowers. The groups are quite vocal, communicating with other groups by singing, roaring and other vocalisations. Before singing, the indri move to the treetops, which allows them to be heard up to almost 2.5 miles (4 km) away.

Like many other species of lemur, indri live in a female-dominant society. The dominant female often will displace males to lower branches and poorer feeding grounds, and is typically the one to lead the group during travel. It is common for groups to move 300–700 m daily, with most distance travelled midsummer in search of fruit. Indris sleep in trees about 10–30 m above ground and typically sleep alone or in pairs.

The indri is revered by the Malagasy people and plays an important part in their myths and legends. The main threats faced by this lemur are habitat destruction and fragmentation due to slash and burn agriculture, fuelwood gathering and logging. It is also hunted. It is critically endangered.

Leaf-nosed snake   *Langaha madagascariensis*

Commonly known as the Madagascar or Malagasy leaf-nosed snake, it is a medium-sized tree dwelling species. It is found in deciduous dry forests and rainforests, often in vegetation 1.5-2 m (5 to 6 ½ feet) above the ground.

Malagasy leaf-nosed snakes can grow up to 1 m (3 feet) in length. Males are brown on top and yellow on their belly with a long tapering snout, while the females are mottled grey with a flattened, leaf shaped snout. The function of their appendage is unknown, but obviously also serves as camouflage.

This snake is largely a sit-and-wait predator. It may show curious resting behavior, hanging straight down from a branch. It eats arboreal and terrestrial lizards. It also exhibits hooding while stalking prey. These hooding and swaying behaviors, along with its coloring, might allow it to mimic a vine swaying in the wind.

Malagasy leaf-nosed snakes are generally calm and reluctant to bite unless provoked. A bite causes severe pain in humans but is not deadly.

Leaf-tailed gecko *Uroplatus phantasticus*
This arboreal species lives in dense tropical forests of northern and central Madagascar. The adhesive scales under their fingers and toes and their strong curved claws enable them to easily move through the trees. Leaf-tailed geckos have large eyes and broad flat, leaf-like tails. Because they have no eyelids, just a transparent covering over their eyes, they use their long, mobile tongues to wipe away any dust or debris that gets into their eyes.

Adults average 90mm (3.5 in) in total length, including the tail. Leaf-tailed geckos can be grey-brown, green-brown or black colored. Leaf-tailed geckos are solitary creatures.

Predators can hardly notice it among the branches and leaves due to its excellent camouflage. Leaf-tailed gecko is even able to flatten the body to reduce its own shadow. Satanic leaf-tailed gecko opens its jaws and exposes its bright-red mouth to frighten the predator when it is cornered. Leaf-tailed geckos can also shed its tail to escape from danger.
The leaf-tailed gecko is egg-laying. Reproduction starts at the beginning of the rainy season when it lays two spherical eggs onto the ground under leaf litter or in the dead leaves of plants. Leaf-tailed geckos do not show parental care. Both eggs and hatchlings are left on their own. Leaf-tailed gecko has an average lifespan of 2 to 9 years in the wild.

Habitat destruction, deforestation, and collection for the pet trade all threaten its existence. Leaf-tailed geckos can only inhabit a very specific environment and cannot tolerate any degradation of its natural habitat. Thus the survival of this leaf-tailed gecko is intrinsically linked to the continued existence of Madagascar's rainforest.

Madagascar bright-eyed frog *Boophis madagascariensis*
This tree frog, 6-10 cm long (2.4-4 in) is one of Madagascar's largest frogs. It is found in the rainforest of the highlands down to the east coast.

During the day, it can be occasionally found in leaf axils of large plants, in bamboo tree holes or similar shelters about 1 m (3 ft) above the ground. Males call out loudly for mates in the shallow water, or at the border of pools and slow-moving brooks, hidden in the vegetation.

Females can lay 400 eggs or more! The eggs are black and can be found in shallow, slow-moving water, in the shade. Brownish tadpoles hatch after one week. Metamorphosing juveniles are tiny. They are light green with dark brown spots on the back and dark brown bands on their limbs. After three months, they resemble the brownish adults. This frog is threatened by habitat loss.

Malagasy kingfisher *Corythornis vintsioides*
This beautiful little bird's natural habitat is subtropical or tropical mangrove forests. It is one of only two kingfishers that live in Madagascar. The Malagasy kingfisher is 13 cm (5.1 in) in length with a weight of 16.5 - 22 g (0.58 - 0.78 oz). It has dark blue upperparts, reddish-brown underparts and a crested blue- and green-barred crown. The bill is black. The sexes are alike. Malagasy kingfishers make their home among the reeds of ponds, streams and wetlands and in tropical mangrove forests. They are fast fliers, skimming the water as their short, rounded wings whir. They perch near a river or stream watching for prey to swim into striking range. Then they spring off the perch to hover for a few seconds, diving their heads in and snatching their prey. They toss small fish into the air and swallow them headfirst. They also feed on aquatic insects, frogs and crustaceans.

They excavate a tunnel in a sandy bank to make a burrow. Within the nesting chamber, the female lays three to four clutches of between three and six eggs. The white, round eggs sit on a bed of fish bones and pellets.

Malthe's green-eared chameleon *Calumma malthe*
This chameleon is found in a few isolated pockets and exclusively in the middle of rainforests in dense vegetation near small rivers where the humidity is high. The male has a funny looking nose appendage. These chameleons spread their occipital lobes to look larger and more impressive, when they are stressed or being aggressive. They also hiss.

It is a common misconception that chameleons can change color to match any color of their environments. All chameleons have a natural color range with which they are born, and this range is determined by species. It is affected by temperature, mood and light. If, for example, the color purple is not within the range of colors to which a particular species can change, then they will never turn purple. This species is threatened by climate change.

Mouse lemur *Microcebus tavaratra*
This mouse lemur is found in northern Madagascar and like all mouse lemurs it is nocturnal. Mouse lemurs have a combined head, body and tail length of less than 27 centimetres (11 in), making them the smallest primates. *Microcebus tavaratra* is a relatively large mouse lemur with a head-body length of 12–14 cm (4.7-5.5 in), a tail length of 15–16 cm (~ 6 in), a total length of 28–29 cm (11 in), and a body weight of 45–77 g (1.6-2.7 oz).

Mouse lemurs are omnivorous; their diets are diverse and include insect secretions, insects, small vertebrates, gum, fruit, flowers, nectar, and also leaves and buds depending on the season. In 1992, there were two known mouse lemur species; by 2016, there were 24! The number of species is uncertain and is increasing as more animals are discovered and studied. Many mouse lemur species are endangered because of habitat destruction.

Panther chameleon *Furcifer pardalis*
The panther chameleon is found in the eastern and northern parts of Madagascar in tropical forest. Its color varies with location, and the different color patterns of panther chameleons are commonly referred to as 'locales' which are named after the geographical location where they are found. On each foot, their five toes are fused into a group of two and a group of three, giving the foot a tongs-like appearance. These specialized feet allow the panther chameleon a tight grip on narrow branches. Each toe has a sharp claw to gain traction on surfaces, such as bark, when climbing.

Their eyes are most distinctive. The upper and lower eyelids are joined, with only a pinhole large enough for the pupil to see through. They can rotate and focus their eyes separately to observe two different objects simultaneously! In effect, it gives them 360-degree vision around their bodies. When prey is located, both eyes can be focused in the same direction, giving sharp stereoscopic vision and depth perception. They have keen eyesight and can see small insects 16-32 feet (5–10m) away.

Panther chameleons have very long tongues (sometimes longer than their body), which they can rapidly extend out of their mouth. The tongue hits the prey in about 0.0030 seconds. At the tip of this elastic tongue, a muscular, club-like structure covered in thick mucus forms a suction cup. Once the tip sticks to its prey, it is drawn quickly back into the mouth, where the panther chameleon's strong jaws crush it to eat.

Red-legged golden orb-weaver *Nephila inaurata*
This spider is an extremely large, classic orb weaver, notable for its strength and its characteristic golden color. It has long, slender red legs and a highly elongated, silvery and black abdomen which may have prominent yellow markings. Females may be 20 times larger than adult males, which have more rounded abdomens with darker, black and red markings. They eat flies, mosquitoes, moths, wasps and unfortunate beetles who happen to get tangled up in their webs.

Webs are typically found up to 6 m (18 feet) above the ground, attached to trees, shrubs or poles; normally several are strung together to form enormous "homes" in order to cover a large surface area. Golden orb weavers can occur in large aggregations, with numerous adjoining webs holding hundreds or even thousands of individuals. Like other spiders in its subfamily, it weaves webs so strong that sometimes even birds and bats get caught. Only female spiders construct webs; males occupy a female's web.

Egg sacs the size of a small marble are made of thick silk and contain 100-200 eggs which hatch after two months. This species is commonly kept in captivity.

Red ruffed lemur *Varecia rubra*
Red-ruffed lemurs live only in the northeastern part of Madagascar in deciduous forests. Red ruffed lemurs are entirely vegetarian. They eat some leaves, seeds, and nectar, but most of their diet consists of fruit.

It is one of the largest primates of Madagascar with a body length of 53 cm (21 in), a tail length of 60 cm (24 in) and weigh 3.3–3.6 kg (7.3-8 lbs). Their fur is mainly rusty red in color; however, fur on the forehead, stomach, tail, and inside of limbs is black and they have a white patch on the back of the neck.

Breeding occurs from May until July. After a 90 to 120 day gestation period, females usually give birth to three young (although they can have up to six). Ruffed lemurs are one of only a few primates that have litters of offspring, and females have six nipples so that they can nurse all of their young simultaneously. Infant red ruffed lemurs are not as well developed at birth as other lemurs. Newborns have fur and can see, but as they cannot move, the female leaves them in the nest until they are seven weeks old. The young lemurs stay in the nest while adults search for food, unlike most primates who carry their offspring with them.

Red-ruffed lemurs warn other group members about predators through a number of alarm calls, which appear to vary depending on the location of the predator. Red-ruffed lemurs have a lifespan of 15 to 20 years in the wild. They are critically endangered. Logging, burning of habitat, cyclones, mining, hunting, and the illegal pet trade are primary threats.

Ring-tailed lemur *Lemur catta*
The ring-tailed lemur is a large primate and the most recognized lemur due to its long, black and white ringed tail. It Inhabits gallery forests to spiny scrub in the southern regions of Madagascar. It is active only in daylight hours. The ring-tailed lemur is an opportunistic omnivore, primarily eating fruits and leaves, particularly those of the tamarind tree. When available, tamarind makes up as much as 50% of their diet, especially during the dry, winter season. The ring-tailed lemur eats from as many as three dozen different plant species, and its diet includes flowers, herbs, bark and sap. It has been observed eating decayed wood, earth, spider webs, insect cocoons, spiders, caterpillars, cicadas and grasshoppers, small birds and chameleons.

This lemur is highly social, living in groups of up to 30 individuals. It is also female dominant, a trait common among lemurs. To keep warm and reaffirm social bonds, groups will huddle together. The ring-tailed lemur will also sunbathe, sitting upright facing its underside, with its thinner white fur towards the sun. Like other lemurs, it relies strongly on its sense of smell and marks its territory with scent glands. The males perform a unique scent marking behavior called *spur marking* and will participate in *stink fights* by impregnating their tail with their scent and wafting it at opponents.

Despite reproducing readily in captivity and being the most abundant lemur in zoos worldwide, the ring-tailed lemur is listed as endangered due to habitat destruction, hunting for bushmeat and the exotic pet trade. As of early 2017, the population in the wild is believed to have crashed as low as 2,000 individuals due to habitat loss, poaching and hunting, making them far more critically endangered

Silky sikafa   *Propithecus candidus*
This large lemur is characterized by long, silky, white fur. It is only found within a few protected areas in the rainforests of northeastern Madagascar. The silky sikafa is one of the rarest mammals on Earth, and is listed by the International Union for Conservation of Nature as one of the world's 25 most critically endangered primates.

The silky sifaka lives in groups of two to nine individuals. It spends much of its day feeding and resting, though it also devotes considerable time to social behaviors, such as playing and grooming, as well as travelling. Like other sifakas, it eats mainly leaves and seeds, but also fruit, flowers, and even soil on occasion. It only mates one day a year during the start of the rainy season. As with other sifaka species, group members often groom, play with, occasionally carry, and even nurse infants that are not their own.

The silky sifaka vocalizes frequently despite its moderately sized repertoire consisting of seven adult calls. Like all other lemurs, it relies strongly on scent for communication. Males frequently scent-mark on top of scent marks made by other group members, particularly females. Males also gouge trees with their toothcomb (a special arrangement of the bottom, front teeth) prior to chest scent-marking. Chest marking results in males having brown-stained chests, the only visible trait that can be used to distinguish adult males from adult females.

The silky sifaka is hunted throughout its range as no local fady (taboo) exists against eating it. Habitat disturbance, such as slash-and-burn agriculture (tavy), illegal logging of precious woods (particularly rosewood) and fuelwood, also occurs within the protected areas where it is found.

Tomato frog   *Dyscophus antongilii*
Tomato frogs are found in the northeast of Madagascar. Females are much larger than males, reaching up to 10.5 cm (~ 4 in) and 230 g in weight (~8 oz), compared to 6.5 cm and 41 g for males. Tomato frogs have a vibrant, orange-red color. It is thought that the brilliant colors of the tomato frog act as a warning to potential predators that these frogs are toxic. A white glue-like substance is secreted from its skin to deter predators and the secretion can produce an allergic reaction in humans. The tomato frog breeds in shallow pools, swamps and areas of slow-moving water. It is classified as near threatened.

CREDITS

Thanks to all these terrific artists!

Albert, Amanda - Illustrations 1,29 - Graphic designer and freelance artist living in Tallahassee, FL who has been drawing since she was 2 ½. See more work at https://www.redbubble.com/people/beawsmmakestuff/shop#profile.

Chanaud, Chris - Illustrations 2,13,26,27,28 - This super supportive artist is a good friend of our director. Learn more at https://www.instagram.com/chris_chanaud_art/ or contact at Powermacdaddy@gmail.com.

Choice, Isabella – Illustration 14 - Student at Maclay School, Tallahassee | 12th Grade, Graphic Design Focus, AP 3D Track. She is an advocate for all things representing love and unity in her body of art work.

Culbertson, Kat - Illustrations 36,37 - Peace Corps volunteer who worked with the Green Again team in Ampasimbe, Madagascar. Kat plans to pursue a PhD in tropical forest restoration ecology. You can follow her adventures on her blog site at https://katsnotinkansasanymore.home.blog/.

Dickinson, John-Henry - Illustration 22 - Student at Maclay School, Tallahassee | 10th Grade, Graphic Design Focus with an interest in animation.

Gooch, Ashlynn - Illustration 7 - Student at Maclay School, Tallahassee | 11th Grade, Graphic Design Focus, and Photography Track.

Gray, Stephan - Illustration 4 - Student at Maclay School, Tallahassee | 11th Grade, Graphic Design Focus and passion for illustration in mythological creations.

Krupp, Emily - Illustrations 31,32,33,34,35 - Former Peace Corps volunteer in Fazihay, Madagascar.

Lyons, Katie - Illustrations 6,16 - Student at Maclay School, Tallahassee | 12th Grade, Graphic Design Focus, with an interest in Biology.

Moran, Patrick - Artist/Animator who contributed the AMAZING cover art and worked tirelessly on book assembly. See more work at artofmoran.com.

Palacino, Taylor - Illustrations 3,20 - Mixed media artist who draws on illustrative styles. Taylor wrote "This coloring book project was the ideal outlet to advocate for Green Again's work while broadening artistic abilities." Find more on Instagram @tapalacino or to email directly at palacinota1093@gmail.com.

Ribich, Stephanie - Illustration 21 - Student at Maclay School, Tallahassee | 12th Grade, Graphic Design Focus, AP 2D, AP 3D, Photography Track, and pursuit in the Design Collaboration. Stephanie's work vocalizes the importance of protecting our environment from the destruction of humanization.

Ritson, Lea – Illustrations 8,9,10,12,17,18,19,24 - Former Peace Corps volunteer in Madagascar. Lea participated in a range of projects focusing on agriculture and the environment. Lea uses her creative talents as tools to ignite passion, curiosity, and love for the natural world.

Roland & Pascal - Illustrations 5,11,15,23,30 - Malagasy resident collaborator. These young men just showed up at a coffee shop with art for Matt after hearing he was looking for coloring book art!

Sharkey, Nathan - Illustration 25 - Student at Maclay School, Tallahassee |12th Grade, Graphic Design Focus with a Photography Track and passion for the environment.

More Green Again History!

Green Again Madagascar was founded in 2014 in response to rampant loss of tropical rainforest in northeastern Madagascar. Indeed, the country, as a whole, has lost almost 90% of its rainforest.

Green Again is different from other tree-planting organizations. Rainforest restoration is more than simply planting trees. It involves a long-term commitment in assisting the recovery of an ecosystem. So far, we've planted 49,374 trees of 66 Malagasy species.

A core part of our work is making scientific observations on the growth and survival of each tree that we plant. The data we gather is used for continual improvement of our techniques. We believe in SCIENCE!

Money goes a long way in the developing world, where goods and services are typically much more cost effective than in the developed world. This is especially true in Madagascar, one of the poorest countries in the world.

One US dollar can accomplish in Madagascar
what $100 USD can accomplish in the US.

Contributions to Green Again help buy trees, train and provide income to Malagasy villagers and restore habitat for magnificent, mysterious – and often endangered - animals. We have tree seedling nurseries (and planting operations) in and around four villages and hope to expand.

Please consider making a gift to Green Again. Thank you!

GreenAgain is a 501(c)3 non-profit
Donations are deductible to the full extent allowed by the law!
EIN 83-3651734

Green Again Restoration
P.O. Box 4362
Saint Paul, Minnesota 55104
https://www.greenagainmadagascar.org/
Follow us on Facebook!

Please watch for Coloring Book 2!
More Magnificent, Mysterious Rainforest Animals of Madagascar!

Made in the USA
Columbia, SC
11 December 2020